CAMBRIDGE LIBRARY COLLECTION

Books of enduring scholarly value

Technology

The focus of this series is engineering, broadly construed. It covers technological innovation from a range of periods and cultures, but centres on the technological achievements of the industrial era in the West, particularly in the nineteenth century, as understood by their contemporaries. Infrastructure is one major focus, covering the building of railways and canals, bridges and tunnels, land drainage, the laying of submarine cables, and the construction of docks and lighthouses. Other key topics include developments in industrial and manufacturing fields such as mining technology, the production of iron and steel, the use of steam power, and chemical processes such as photography and textile dyes.

History of the Suez Canal

In the early 1850s the French diplomat and engineer Ferdinand de Lesseps (1805–94) revived earlier French plans to build a canal through the Isthmus of Suez, and, thanks to his good relations with the Viceroy of Egypt, won approval for the project in the face of British and Turkish opposition. This 1870 lecture reveals De Lesseps' enchantment with the desert and its people, his determination to complete the canal, and his annoyance at British antagonism. By 1875, when this English translation by Sir Henry Wolff was published, the canal had been open for six years and the British position had shifted dramatically. The government bought Egypt's shares in the Canal Company, and Wolff was chosen by Disraeli to speak in Parliament in support of the purchase. De Lesseps' book remains an invaluable source on the canal, the politics of the major powers, and European attitudes towards the Middle East.

Cambridge University Press has long been a pioneer in the reissuing of out-of-print titles from its own backlist, producing digital reprints of books that are still sought after by scholars and students but could not be reprinted economically using traditional technology. The Cambridge Library Collection extends this activity to a wider range of books which are still of importance to researchers and professionals, either for the source material they contain, or as landmarks in the history of their academic discipline.

Drawing from the world-renowned collections in the Cambridge University Library, and guided by the advice of experts in each subject area, Cambridge University Press is using state-of-the-art scanning machines in its own Printing House to capture the content of each book selected for inclusion. The files are processed to give a consistently clear, crisp image, and the books finished to the high quality standard for which the Press is recognised around the world. The latest print-on-demand technology ensures that the books will remain available indefinitely, and that orders for single or multiple copies can quickly be supplied.

The Cambridge Library Collection will bring back to life books of enduring scholarly value (including out-of-copyright works originally issued by other publishers) across a wide range of disciplines in the humanities and social sciences and in science and technology.

History of the Suez Canal

A Personal Narrative

Ferdinand de Lesseps

CAMBRIDGE UNIVERSITY PRESS

Cambridge, New York, Melbourne, Madrid, Cape Town, Singapore,
São Paolo, Delhi, Dubai, Tokyo, Mexico City

Published in the United States of America by Cambridge University Press, New York

www.cambridge.org
Information on this title: www.cambridge.org/9781108026383

This edition first published 1876
This digitally printed version 2011

ISBN 978-1-108-02638-3 Paperback

THE

HISTORY OF THE SUEZ CANAL

"J'AI POUR PRINCIPE DE COMMENCER PAR AVOIR DE LA CONFIANCE."—*Monsieur de Lesseps.*

THE

HISTORY OF THE SUEZ CANAL

A PERSONAL NARRATIVÉ

BY

MONSIEUR FERDINAND DE LESSEPS

G.C.S.I.

TRANSLATED FROM THE FRENCH, BY PERMISSION OF THE AUTHOR, BY

SIR HENRY DRUMMOND WOLFF

K.C.M.G.

M.P. FOR CHRISTCHURCH

WILLIAM BLACKWOOD AND SONS

EDINBURGH AND LONDON

MDCCCLXXVI

PREFACE.

THE following is the translation of a lecture given by Monsieur de Lesseps in April 1870, before the Société des Gens de Lettres at Paris. It was taken down in shorthand by Monsieur Sabbatier, stenographer of the Corps Legislatif. Such as it is printed in French I have endeavoured to reproduce it in English without addition or retrenchment. As it stands, the text gives a lively picture of the origin and progress of the Canal to its completion. It also gives the unbiassed views of the projector of the Canal as he wished to lay them before his audience—

and more, it presents a vivid picture of the author of the Canal himself, with that hopefulness and decision, courage, resource, and caution, so genially blended in his enthusiastic and steadfast nature.

At a period when so splendid a moral reparation has been made to him, I have not thought it necessary to erase some expressions of natural irritation at the opposition shown to his scheme originally by the British Government. But it is worthy of note that he has always drawn a distinction between the reception given to his project by the British Government and that accorded him by the people; and I believe no one is more frankly pleased than himself at a policy, however tardy, which gives to his scheme the international character put forward in the very title of his Company, "Compagnie Universelle du Canal Maritime de Suez."

The *bona fides* of these views is sufficiently

established by the terms of the concession given by the Viceroy on January 5, 1856.

It is there laid down :—

Article 14. We solemnly declare, for ourselves and our successors, subject to the ratification of his Imperial Majesty the Sultan, the Grand Maritime Canal from Suez to Pelusium and the ports dependent thereon open always, as a neutral passage, to every commercial vessel crossing from one sea to the other, without any distinction, exclusion, or preference of persons or nationalities. . . .

Article 15 says: In consequence of the principle laid down in the preceding article, the Universal Concessionary Company cannot, in any case, grant to any ship, company, or person, any advantages or power not granted to other ships, &c., &c.

In the statutes it is stipulated :—

Article 7. The share certificates are to

be printed in Turkish, German, English, French, and Italian.

Article 8. The shares are to be subscribed and paid for at Alexandria, Amsterdam, Constantinople, London, New York, Paris, St Petersburg, Vienna, Genoa, Barcelona, &c.

By Article 24 a Council of Administration is formed, composed of twenty-one members, representing the principal nationalities interested in the undertaking.

But while, no doubt, the opposition of Lord Palmerston was unfortunate, it was to some extent justified by circumstances that have since ceased to exist. The grants of land given up to a French company, governed by French laws, and having its seat at Paris, did point to a permanent French territorial settlement in Egypt, inconsistent with a real neutrality, and likely to lead to grave political difficulty; while the opinion of Mr Stephenson, given openly in Parliament, cast doubts on the commercial prospects of the enterprise. The latter opposition is now controverted by facts; while an arrangement entered into previous to the opening of the Canal, did away with the former apprehensions.

The recent act of our Government seems to pave the way to a realisation of Monsieur de Lesseps' original idea, by giving us a *locus standi* in any future international negotiation.

There is one question consequent on the purchase of the shares by our Government, which will require serious and early consideration. This is the necessity of further expenditure, for the purpose of "rendering more easy and more rapid the passage of the Canal." These words are quoted from the report of the Council of Administration of 1875. England will now have to take a new attitude in the discussion of this problem. Hitherto she has been only a customer—now she has become a partner.

While considering this question, it will be necessary and not uninteresting to trace the progress of the traffic through the Canal since the opening. This will be seen at a glance from the following table :—

Year.	Number of Vessels.	Gross tonnage.	Receipts.
1870.	486	654,915$\frac{020}{1000}$	£206,373
1871.	765	1,142,200$\frac{460}{1000}$	359,748
1872.	1082	1,744,481$\frac{320}{1000}$	656,303
1873.	1173	2,085,072$\frac{615}{1000}$	915,892
1874.	1264	2,423,672$\frac{229}{1000}$	994,375

During the first six months of 1875, 822 vessels, of a capacity of 1,546,060$\frac{067}{1000}$, passed the Canal, producing a return of £610,730.

The report points out that these figures give an increase, for the first six months of 1874, of 28 per cent on the number of ships, of 31 per cent on tonnage, and of 21 per cent on receipts. The increase of receipts would have been greater but for the diminution of tolls, of which M. Lesseps complains bitterly. It is unnecessary at present to enter on this question. But the point to which attention should be called is the necessity which is pressing for works for the improvement and enlargement of the Canal—works which must continually be increased for a traffic augmenting with an incredible momentum. The tonnage—the real gauge of material requirement—has quadrupled in five years. On the 3d of March last, thirty-three ships were at the same time passing the Canal. New companies are being formed, and existing companies are adding to their regular service. The detailed statements given by M. de Lesseps in his report are almost beyond belief. It is perfectly plain that ere long the present dimensions of the Canal will be inadequate to the requirements of the two hemispheres. M. de Lesseps discovered in his early studies that the traffic from East to West doubled every ten years. The Suez Canal has quadrupled its own traffic in five years.

To my mind the recent action of the Government is an event almost as great as the construction of the Canal. It inaugurates not only a new policy, but a new scheme of policy, and opens out a field for a full understanding between the different nations interested in the trade between East and West. The complete redemption of the Canal from the hands of the Company, and the purchase of the Viceroy's reversionary interest, will be now far easier than before; and to this Lord Derby's language, reproduced in the French yellow-book, seems to point.

BOSCOMBE TOWER, BOURNEMOUTH, HANTS,
December 18, 1875.

HISTORY

OF

THE SUEZ CANAL.

Ladies and Gentlemen,—

I have eagerly availed myself of the amiable invitation of my colleagues in the Society of the Men of Letters; and it is always with great pleasure that I return to the neighbourhood of the schools. I cannot forget that at the School of Medicine I for the first time had the honour of addressing the public on the Suez Canal. I began with our patriotic and impetuous youth;

with youth and woman on one's side, success is certain. (Loud applause.)

In this last lecture I shall be happy to retrace the historic facts of the cutting of the Suez Canal. Everything concerning the negotiations has been published. The conventions with the Egyptian Government are known by all. On the labours of the engineers, M. Lavelley has made reports to the Society of Civil Engineers. These different questions are well understood by the public, which knows the Isthmus of Suez as thoroughly as though that isthmus were in the neighbourhood of Paris. I will content myself with giving you a summary of the circumstances which led to or accompanied the execution of the work. My story will perhaps be useful and serviceable to those who wish to study the connection of facts, and who study the human heart. Nothing is so logical as facts. I will tell them to you without preparation, and as they rise to my memory; choosing only the principal points, or those that to my mind ought to interest you. (Hear, hear.)

I am asked every day how the idea of the Canal occurred to me. Nothing useful is done without cause, without study, and without reflection. An illustrious statesman, Monsieur Guizot, has said that time respects only what itself has made. It was after five years of study and of meditation in my closet, five years of investigation and of preparatory labours in the isthmus, and eleven years of execution, that we attained the end of our efforts.

In 1849 I was sent by the Government on an extraordinary mission to Rome, in conformity with a vote of the sovereign Assembly. I was to follow a line of conduct determined by that vote. When the Legislative replaced the Constituent Assembly I was asked to follow another line of conduct, which it is not for me to blame, but which I could not adopt. Unwilling to betray my mission, I abandoned twenty-nine years of diplomatic service. Being thus released from politics, I gave myself up to my first studies of the East and Egypt, all the while creating a farm in Berry. This state of things lasted some time. Many cast the stone at me during this period, and turned against me, reproaching me for not having changed my opinions and conduct. Events have shown, I think, that the policy opposed to that which I was ordered to pursue, and which was in harmony with my ideas, was not fortunate for the interests of our country.

On applying myself to the study of oriental questions, my mind naturally turned to the Isthmus of Suez. Every intelligent child, on first seeing the map, must have asked his teacher why the road to India was not across the Isthmus of Suez. The master answered that there was a difference of level between the Red Sea and the Mediterranean—that it was impossible to dig out in the desert a canal which should not at once be filled up by sand, &c., &c.

But now all these phantoms have disappeared. What was impossible fifty years ago has become easy with steam, the electric telegraph, and all the appliances which science has placed at our disposal.

From 1849 to 1854 I studied everything connected with the trade between the West and the East. I discovered that the traffic doubled every ten years, and that the time had arrived at which the formation of a company for the construction of the Suez Canal could develop that traffic in a marvellous manner. In 1852, when my studies were completed, and I found before me the alternative either of enlisting in my cause a Viceroy of Egypt absorbed in pleasure, or of applying to Constantinople, I took the latter course. My family and friendly connections caused my application to be examined, and obtained for me the answer that the solution of the question in no wise concerned the Porte. Observe, that later, when Egypt had taken the initiative in the Canal, England, which had, without the intervention of the Divan, obtained the construction of the railway between Alexandria and Suez, remonstrated with the Porte in the name of its ignored rights. I then kept back my scheme, and devoted my attention to my cattle and my farm. (Laughter.)

One day, while on the roof of a house I was building, in the midst of scaffolding and carpenters, I received a newspaper which announced the death of the Pacha, and the accession of Mohammed Saïd, son of Mehemet Ali.

While residing as the French agent accredited to Mehemet Ali, that great prince had shown me much affection on account of the memory of my father, who, when representing France in Egypt after the peace of Amiens, had contributed to the elevation of the Bim-bachi Mehemet-Ali-Aga, who had recently arrived from Macedonia with a contingent of a thousand men.

The First Consul, Bonaparte, and the Prince de Talleyrand, minister of external relations, had instructed their agent to seek amongst the Turkish militia for a bold and intelligent man to be named from Constantinople Pacha of Cairo, a title almost nominal, and who could serve to break down the power of the Mamelukes, who were hostile to French policy. One of my father's janissaries brought to him one day Mehemet-

Ali-Aga, who at that period could neither read nor write. He had left Kavalla with his little band, and sometimes boasted of coming from the same country as Alexander. Thirty years later, when the consular corps came to Alexandria to compliment Mehemet Ali Pacha on the victories of his son Ibrahim Pacha in Syria, the Viceroy of Egypt, turning towards me, said to my colleague: "The father of this young man was a great personage when I was a very small one. He had one day invited me to dinner. The next day I learnt that some silver had been stolen from his table, and as I was the only person who could be suspected of the theft, I dared not return to the house of the French agent, who was obliged to send for me and reassure me." This was very fine from the lips of a triumphant man, avowing that he might plausibly have been accused of theft. (Laughter.) Such was the origin of my relations with Egypt and the family of Mehemet Ali, and consequently of my friendship with Saïd Pacha. His father was an extremely severe man, who was annoyed at seeing him grow fat to a formidable extent—(renewed laughter)—and who, to prevent excessive obesity in a child he loved, sent him to climb the masts of ships for two hours a-day, to skip with a rope, to row, and to walk round the walls of the city. I was at that time the only person authorised to receive him. When he came to me he would throw himself on my divan quite worn out. He had come to an understanding with my servants, as he confessed to me later, to obtain from them secretly meals of macaroni, to make up for the fasting imposed on him. The Prince was brought up in French ideas with an impetuous head and great sincerity of character.

When Saïd Pacha succeeded to power it

was my first care to congratulate him. Two years before he had been accused of conspiracy. While a conspiracy is going on no one confesses to being a party to it. He had been ill treated by the Viceroy. His family had been exiled. The discontented had gathered round him, and . . . he had been obliged to escape as he could. He came to Paris, and lived at a hotel in the Rue de Richelieu, where I visited him. His situation, the welcome I gave him, and the recollections of his childhood, established between us from that moment a truly brotherly friendship. Shortly after he returned to Egypt, and when in 1854 he was called to succeed Abbas Pacha, he fixed a meeting for me at Alexandria for November 1854. I went there. He gave me one of his palaces as a residence, and invited me to accompany him to Cairo, crossing the

Libyan desert with a little army of eleven thousand men.

The Viceroy pitched his camp on the ruins of Marea, beyond Lake Mareotis. I went to join him. I had always carried my project in my head, but I awaited the favourable moment to speak of it, for I wished first to acquaint the Prince with the system, new for him, of limited financial associations, which can bring capital to a country without depriving a sovereign of his influence—assisting him, on the contrary, to increase his power, by means intended to advance the public prosperity. It was further essential to conciliate the goodwill of the Viceroy's intimate associates, consisting principally of the old councillors of his father, who were more skilful in the exercise of the horse than of the brain. I used to ride with them in the desert, my talent for riding having conquered their esteem. Intimate with Saïd's old companion from childhood, his minister, Zulfikar Pacha, who had been brought up in the French school, and could thus understand everything, I initiated him into my project, and it was agreed that he should acquaint me when the day arrived which he thought opportune for me to speak on the subject to his master.

Two months passed, and on the day named, the 30th of November 1854, I presented myself at the tent of the Viceroy, placed on an eminence surrounded by a wall of rough stones, forming a little fortification with embrasures for cannon. I had remarked that there was a place where one could leap with a horse over the parapet, there being a terrace outside on which the horse had the chance of a footing.

The Viceroy welcomed my project, and requested me to go to my tent to prepare a report for him, which he permitted me to bring him. His councillors and generals were around him. I vaulted on my horse, which leaped the parapet, galloped down the slope, and then brought me back to the enclosure when I had taken the time necessary to draw up the report, which had been ready for several years. The whole question was clearly set forth in a page and a half; and when the Prince himself had read it to his followers, accompanying it with a translation in Turkish, and had asked their advice, he received the unanimous answer that the proposal of the guest, whose friendship for the family of Mehemet Ali was known, could not be otherwise than favourable, and that it was desirable to accept it.

The concession was immediately granted. The word of Mohammed Saïd was as good as a contract.

On arriving at Cairo he received in front of the citadel the representatives of the different Governments who came to congratulate him on his accession to the viceroyalty. He then said to the Consul-General of America—"I shall queen the pawn against you Americans. The Isthmus of Suez will be pierced before yours."

He then continued to speak of the project. The Consul-General of England seemed agitated. Being present at the audience, and on a sign of the Prince, I remarked that the enterprise, as then conceived, ought to offend no one—that all countries would contribute to it equally, if they desired, by a public subscription, and that if I were charged with the formation of a financial company for carrying it out, it was not as a

Frenchman, but as a friend of Egypt and the Viceroy. Each Consul-General hastened to transmit the news to his Government, and the answer was the despatch to Mohammed Saïd of the Grand Cross of the Orders of nearly all the sovereigns. (Hear, hear.)

The act of concession was legally executed on the 30th of November 1854. An excursion was decided on for the exploration of the isthmus. The Viceroy associated with me three French engineers in his service —Messrs Mougel Bey, Linant Bey, and Aïvas. Four persons required at least sixty camels, of which twenty-five were loaded with water, to cross a desert now peopled by 40,000 inhabitants. We left Cairo; we crossed the isthmus from north to south, studying the nature of the land, examining the possibility of a fresh track —for from the most remote times attention had been directed only to an inland canal from the Nile to the Red Sea, and never to a canal without locks dug directly from sea to sea. The scheme of a fluvial and not a maritime canal was that adopted by the Saint Simonians and by Father Enfantin, to whom we owe the studies of 1847, and the recognition of the even level of the two seas.

Former projects, including that of M. Lepere, the engineer in chief of the French expedition to Egypt, had employed the water of the Nile for the navigation of the canal by means of channels and sluices. This was a mistake, and this will make it impossible for the American plans for the cutting of the Isthmus of Panama to succeed, until they find a means of simply piercing the isthmus from sea to sea. No one will ever succeed in making a maritime canal by bringing water from an inland river to the sea.

Moreover, for a passage which will shorten the journey by three thousand miles, a time will necessarily arrive when you will have perhaps one hundred vessels a-day. The passage for each will require at least half an hour, and there are not a hundred half hours in the day. Then the locks are a human work which must be kept up and repaired. Hence forced delays, a large consumption of water, and no absolute certainty. I think that none of the actual American schemes can lead to good results. I say so here before representatives of America. They must be persuaded that there is no difference of level between the Pacific and Atlantic Oceans, except the difference in the height of the tides on the coasts. Laplace and Fourier denied it for fifty years before all the Academies. (Hear, hear.) We have perfectly proved that there is no difference between the Red Sea and the Mediterranean, except such as is caused by the tides. In America the same. I say it loudly. The Americans can only succeed after serious study of this question. They have traced their projects in red and blue lines on the map without making soundings or levels, or any of the works which preceded our undertaking. We passed five years in the desert, and there made all the preliminary studies before appealing for capital, and we only formed the company to carry them out after receiving the verdict of European science.

Let the good people who are engaged in the American isthmus also make these long and necessary preparatory studies. The Suez Canal was made, thanks to the co-operation of the superior and competent men whom we called in. They made an estimate which was not exceeded in the works by a single centime—be it well understood. Science carried the day on every point. (Hear, hear.)

Our first exploration was long and difficult, and the final result was that to which my instinct had led me,—that is to say, that we were not to make use of the water of the Nile for the navigation of the Suez Canal. During our journey the feet of our camels trampled on the salt crusts of the Bitter Lakes. The lakes are forty leagues in circumference, and are evidently the ancient gulf of Heroöpolis. It was through this desert, converted into an inland sea, that on the day of inauguration on the 17th of November last, a fleet passed—the Aigle at the head. (Applause.)

This basin now contains two thousand million metres of water (440,000,000,000 gallons). In 1854, our caravan in crossing it carried our water, our victuals, sheep, and fowls. Beyond these animals there was not even a fly in this hideous desert. At night we opened the cages of our fowls, full of confidence, for we were sure that the next morning all our beasts would come round us, not to be abandoned in these desolate places where solitude is death. When we struck our camp of a morning, if at the moment of departure a hen had lurked behind pecking at the foot of a tamarisk shrub, quick she would jump up frightened on the back of a camel to regain her cage. (Laughter.) The Fellahs whom I had brought were in constant anxiety, for the inhabitants of the borders of the Nile have the greatest fear of the desert. Well, it is this desert that we explored in every direction for two months in December 1854 and January 1855. We experienced storms, but I must say that these sands of the deserts do not produce the serious inconveniences attributed to them. They are less annoying than the rain and hail which in our climate surprise us in our walks. I have traversed the deserts of

Africa nearly to the equator; I have travelled 350 leagues, mounted on a dromedary, in the season of the south winds, and I have never been stopped by these winds, said to be so violent, even when they blew straight in my face.

One of our travelling companions said that such was the penetrativeness of the sand that it entered almost into the cases of watches closed hermetically. One day, when the wind came of a sudden during breakfast, we wrapped ourselves in our cloaks to take our meal quietly. This engineer, persuaded that the sand would penetrate through the slightest fissure, sought to cover himself over. But, without remarking it, he had left a hole over his head, through which I amused myself by pouring sand. (Laughter.) "See," he said to me, "the sand has even forced its way through cloth." We are constantly threatened with the invasion of the sands in the Canal, and the impossibility of freeing ourselves from them. This prejudice is so deeply rooted in the public mind, that every day it is alleged as a formidable obstacle to the maintenance of the Canal.

After the passage of 130 vessels during the *fêtes* of inauguration, no deposit of sand, no slip, was reported. Since that period two, three, four, and five ships have passed daily, and the Canal is as intact as before the inauguration. (Hear, hear, and applause.)

I received, last evening, a telegraphic despatch which announces that during the month of March we had taken 640,000 francs, and that six ships had passed since the 7th, which makes twenty-two since the 1st of the month.* (Renewed applause.)

The progress is this. I think it right to

* The following is the traffic reported for 1874 :—

inform you of it, though interrupting the order of my ideas, to show you the ascending scale of the traffic. The vessels which passed the Canal were, in number, nine in December, nineteen in January, twenty-nine in February, fifty-two in March, &c. From the beginning to the 9th of this month (April) we already count twenty-two steam vessels.* (Prolonged cheers.)

You see that steam has superseded sails. I apologise to the sailing vessels, which, however, will find a last refuge in the much-calumniated Red Sea. Improvements have been invented for steamers which consider-

	Tons.
English,	1,797,000
French,	222,000
Dutch,	103,000
Austrian,	84,000
Italian,	63,000
Spanish,	50,000
German,	39,000
Various,	65,000

* See Preface.

ably reduce the space formerly taken up by the machinery, and which insure a saving of 50 per cent on the consumption of coal. The English steamer Brazilian, sailing from Bombay, arrived at Liverpool, carrying in its hold 13,000 bales of cotton and 2500 bales of wool, equal to 4000 tons. And again—and this is an admirable example of the encouragement given by England to commerce — another vessel, sailing from Bombay, passes the Canal, and leaves its cargo of cotton on the quays of Liverpool. The cotton, immediately sent to Manchester, is manufactured; and nine days later, the ship, with its former cargo in a manufactured state, again sets sail, and returns to India by the Canal. Thus it has been found possible, in seventy days, to bring the raw cotton from India, to unload in England, and to send it back manufactured to India. (Applause.) I wish to contrast this example of devouring activity with the desert once so arid where our hens were so afraid of being forgotten. Now the desert is peopled. We have there three important towns. The period of its first appearance deserves comparison with the present date. (Hear.)

After this digression permit me to return to my narrative. After we had accomplished our first exploration, and the engineers of the Viceroy had drawn up their preliminary plan, I repaired to Constantinople to prepare the execution of the project, so as not to be accused of too much impatience. Though often bold and enterprising, I am happy to show that I could be patient when necessary. I have never endangered anything. My ardour has often been attacked; but in all circumstances I have acted cautiously, and, above all, I have never failed to follow the straight road, which is the only one that leads with certainty to success. Armed with truth, one is always sure of victory. (Applause.) I therefore repaired to Constantinople at the time of the Crimean war. England being opposed to the Canal, I came to an understanding with the Sultan so as to avoid all collision between the two policies. I contented myself with a vizierial letter addressed to the Viceroy, permitting the latter to continue his interest in the Canal. On arriving in Egypt, I gave this letter to the Viceroy, who was much satisfied with it. We organised all the preparations for the studies, and it was decided that I should address myself for the completion of them to the most skilful engineers in Europe.

I had to struggle some time on my return to France against the partisans of the indirect track. I stood alone, without any connection with the press, against scientific men of great merit.

I adopted the plan of answering science by science. I wrote to the Ministers of the principal Powers to designate the engineers holding the first rank in their country; and I asked that they should be authorised to join us.

Austria gave us M. de Negrelli; Italy, M. Paleocapa; Spain, M. Montesino; Holland, M. Conrad, Director-general of the Water Service; Prussia, M. Lentzé, sent by M. Humboldt. As in England there is no recognised corps of engineers, I visited that country, and chose Messrs Rendel, Maclean, and Manby, distinguished engineers; and also a seaman, Captain Harris, who had made seventy voyages in the Red Sea.

France placed at our disposal M. Renaud, Inspector-General of Public Works (*ponts et chausseés*); M. Lieussou, hydrographic engineer of the navy; and Admirals Rigault de Genouilly and Jaurès.

This congress of learned men was convoked by a private gentleman, to meet at Paris on a third storey in the Rue Richepance.*

Most of the engineers were unacquainted with each other. They were the most competent men to be found, who together presented the greatest amount of practical knowledge. They had left their business, the direction of their works, with remarkable disinterestedness, to found the era of a new civilisation. On the day fixed, at eight o'clock in the morning, they were all punctual, arriving by railway from Madrid, Amsterdam, Berlin, Vienna, and London. After introduction, we held our first sitting, at the close of which I could no longer doubt of the success of my enterprise. You may well believe, gentlemen, the congress of these distinguished men did not take place in a moneyed inter-

* Monsieur de Lesseps' present residence.

est. No. Not one of these *savants* would even accept the repayment of his travelling expenses. (Applause.) They named a sub-committee, charged with the study of the land in Egypt. This sub-commission, composed of five members, achieved its task, in the midst of every difficulty, with indefatigable zeal and devotion. On arriving in Egypt, it travelled over the whole of Upper Egypt. On leaving, the Viceroy waited for them at the barrage of the Nile. Sovereigns love to play at soldiers. (Laughter.) The Viceroy, who had his troops around him, in full dress, received the members of the Commission with the highest honours.

For these I thanked him. I thanked him above all for having received them as crowned heads. "But," he said, "are they not the crowned heads of science?" He sent for his tutor, and said to us, "I am going to place my tutor near you at table, because it is he that taught me; if I owe something to any one, it is to M. Koenig, for science is above existence. He has often sentenced me to dry bread and water; but I will not now treat him in the same way. He shall breakfast with us." (Approving smiles.)

He generously defrayed from his private chest all the expenses of the explorations, and the studies of the Commission, which travelled as far as the first cataract. These expenses amounted to three hundred thousand francs (£12,000), for which he declined to be reimbursed when the Company was formed four years later. A frigate came to wait for the Commission at Pelusium, and on the 1st of January 1856 we returned to Alexandria, where the Viceroy waited for us at the gates of his palace. When he learned that the Commission considered the canal possible, by channelling the isthmus from sea to sea, without having recourse to the water of the Nile, he threw himself into my arms, and showed the liveliest satisfaction.

He begged me to return to France with the Commission, to publish its report, and to begin a propaganda in England.

I left, furnished with the definitive act of concession, and the statutes of the Company, which was to be formed as soon as I found a fitting opportunity.

During my first journey to England, while finding sympathy in the commercial and lettered classes, I had found heads of wood among the politicians. (Applause.)

They said, as did of old the magicians of Pharaoh, that this work was impossible: that there was a great difference of level between the two seas. Ah! the magicians of antiquity were not otherwise than modern politicians. (Laughter.) There is nothing uncommon in *doctrinaires* being mistaken.

Before going to England I had published a work at Paris to prepare the public mind for the report of the engineers. When in England, I published the same work in English; but I did not as yet hold meetings: I simply explained my plan to some men of business. One day I go to an English publisher. And this is noteworthy. In France we make too much of the stings of the press. In England no one minds them. There nothing stops you. Every one says what he thinks, and the truth is not long hidden; for the majority of mankind is better than is believed, and in the long-run good carries the day against evil. (Applause.)

I therefore go to my English publisher, and tell him it is my wish to make my work known, to circulate it as much as possible, and to get every one to read it.

The publisher promises me an answer for the next day. Next morning I return to him, and he gives me a bill of costs, in which the largest item is intended for an attack on the work. (Laughter.) We must believe that the epidermis of the English is less sensitive than ours. We certainly do not pay for the rods that scourge us. (Renewed laughter.) "There is no need for praise of a book," says the publisher; "when it is attacked, honest people want to see it, and judge for themselves. How many works have had an immense run only because they have been cried down!" The English publisher was a man of good practical sense. On my return to Paris, I published the engineers' report, which made a great sensation.

It was now necessary to return to Egypt to carry out the project—to try soundings at intervals of 150 to 200 metres, and to take levels. The engineers charged with the preparatory studies laboured with intelligence and devotion. Certainly it is not without cause that search is eagerly made in every country of the world for engineers brought up in the Polytechnic School, and that France glories in them. (Hear, hear.)

I arrive in Egypt. As soon as English politicians perceive the favourable tendency of our affairs, their agents omit no means of damaging us, even going so far as to threaten the Viceroy with forfeiture. They even try to make him out a madman. I had been honoured with the same compliment—(laughter)—at the time of my mission to Rome. It is thus people are treated nowadays. A hundred and fifty years ago they would have been shut up in the Bastile. (Sensation.)

I endeavoured to encourage the Viceroy, telling him that he had nothing to fear;

that I had sounded public opinion in England; and that it was on our side. But nothing succeeded. I found him discouraged, ill, and irritated beyond measure. The blood flew to his head. At length he told me one evening that he could no longer resist all these worries; that attempts had been made to tamper with his troops, whose officers were Turks, and to excite them to desertion. I pointed out to him that as nothing which went on in the desert was known to any one, we had only to do the work required by the Commission, and to take an excursion in the Soudan as far as Khartoum. There populations are found which have been decimated, and have suffered for forty years. The elder brother of Mehemet Ali had been sent there at that period. On his arrival he fixed the imposts at 1000 camels, 1000 slaves, 1000 loads of wood. He required everything by the thousand. The inhabitants had no alternative but submission. One day that the Prince, surrounded by his staff, was engaged at a convivial meal, the insurgent chiefs surrounded the camp with a belt of combustibles which formed part of the tribute. The fire formed an immense circle, and every Egyptian who sought to escape was killed by the arrows of the Soudanians. It was a fearful massacre, and it cannot be said to have been undeserved.

Vengeance was confided by Mehemet Ali to his son-in-law, the famous Defderdar, who committed real atrocities in this country. More than a hundred thousand slaves were torn away to be sent to Egypt. The name of this man remains a synonym for the Scourge of God. Would you believe it, that he one day had the barbarity to cause a groom to be shod like a horse for having badly shoed his charger!

A woman of the country one day complained of a soldier who had bought milk of her and refused to pay for it. "Art thou sure of it?" asked the tyrant. "Take care! they will tear open thy stomach if no milk is found in that of the soldier." (Movement of horror.) They opened the stomach of the soldier. Milk was found in it. For forty years these communities had been in a deplorable state. I urged Saïd Pacha to employ his leisure in carrying relief to this great misery, and I promised to accompany him.

We left for Upper Egypt, and traversed the desert of Korosko. On his arrival in Nubia the miserable state of the people afflicted him, for he was very full of feeling.

We had agreed to meet at Berber, the ancient capital of the empire of Meroë, where the cataracts cease. It was the 1st of January 1857, and I was anxious to wish him a "happy new year." I travel thirty leagues in a few hours. I surprise him in his tent and find him crying like a child. "What is the matter?" I ask. "When my generals came just now," he said, "and put the same question, I answered that the music had touched me: but it is really the fate of this unhappy country, whose sorrows have been wrought by my family; and when I think there is no remedy for it, I feel it a great affliction."

He continued to hold meetings in the neighbouring villages, which have all great squares and fortifications, and asked me to accompany him.

One day there were more than 150,000 persons who had come to see him from the very heart of Africa. A really curious thing is the ease with which journeys are undertaken in these countries. While in presence of the crowd it was reported to the

Prince that in spite of his formal prohibition an old Turk had shut up a slave in his cellar. He orders the master to be bastinadoed and put in irons. Finally, not to disappoint the public enthusiasm, he yielded to a fine impulse of generosity. "Go," he said, "take the cannon of the citadel and throw them into the Nile." It is impossible to depict the transport, the excess of joy, such an order excited among this multitude. As for me, I was a little disturbed. "Do you think you are not going a little too far, and that we can always be sure of these people?" I objected to the Viceroy. "The cannon are too old," he answered, "to fire a single shot." (Laughter.) When all were assembled, the Viceroy declared that he would leave the inhabitants to govern themselves; that he would give them no more Turkish chiefs, and that he wished to establish among them municipalities, which, from the beginning of the world, have been the element of all society.

We then proceeded to Khartoum, which means "an elephant's trunk," because the town is situated as though between two tusks, between the Blue River and the White River. Khartoum is placed at the point of junction. It is a town of 40,000 souls, founded by Mehemet Ali. I arrive in the evening at the Viceroy's, who was very gay. He told me laughingly that on his arrival he had been welcomed with military music executed on instruments which the apothecary of the regiment had mended as well as he could with sticking-plaster. But scarcely were we at table when his countenance fell. He again deplored his inability to do anything to remedy the misery of which his family had been the cause, and declared that nothing remained for him but to abandon the country completely.

The education of this Prince was extensive. He knew the Holy Books and the Commentaries of the Koran. We were sitting peacefully when of a sudden he rises, takes his sword and throws it against the wall. He is wild with fury; he requests me to withdraw to my own room. He wished to pass the night in his reception-room. None of his ministers dared go near him. In Egypt when the Viceroy is angry every one runs away. (Laughter.) All night I had in my room the ministers of the Pacha, who thought him mad. We from time to time sent a Bey to find out what he was doing. At three in the morning he asks for a bath. At daybreak he sends for me. I find him on a divan. "Lesseps," he says, "you wanted to travel on the White Nile. I give you leave."

"You were suffering yesterday?" I asked him. "Forgive me," he said. "It was not against you that I was furious, but against myself. I saw the evil, and could not see the remedy. I was irritated at not having had your practical idea of giving laws to this country, and trying to organise it. On your return, you will see that you will be satisfied with me."

I embarked to ascend the White Nile with Arakel Bey, brother of Nubar Pacha,* an amiable and intelligent young man brought up in France, and ambitious of good. We saw arriving from every side, mounted on dromedaries, caravans who wished, as they said, to thank the great Prince who was giving liberty to their country. The report had spread throughout the desert. Nine days later I return to the Viceroy. He tells me that he has promulgated three decrees which, to my mind, are a model of organisation for a

* The present Viceroy's favourite minister—a statesman of great ability and probity.—H. D. W.

new society. The foundations of them are generosity, loyalty, and straightforwardness. (Hear, hear.)

Arakel Bey, named governor-general of the Soudan, was charged with the execution of these ordinances. Unfortunately, his premature death destroyed the hopes founded on his administration.

We had decided to return to Egypt. Instead of travelling by the desert of Korosko, we changed our route, and took the opposite road by the great desert of Bayouda. During this journey of 350 leagues (1050 miles) I always travelled without arms, and never had cause for alarm. Laden with arms, laden with fears, is the saying. (Approving smiles.) I always kept nine days' distance from the Viceroy, on account of the provision of water for our caravans, and I was always well provided with the necessary supplies.

"How was it," the Prince often asked

me, "that you swam in abundance, while we missed everything?" "I quite understand it. Your Government has treated the people so badly that I often have to suffer from the distrust of the inhabitants. I often have to wait an hour—two hours, before their children venture near me." (Laughter.) "The children are always sent forward to reconnoitre. If they hesitate too long, I throw them some little pieces of money, shells, or glass trinkets. They hasten to tell their mothers what they have seen; and then the women come—not generally the youngest." (Renewed laughter.) "They surround me, and ask me why I have made presents to their children? I answer that I am a generous man travelling for my pleasure and for the good of the countries I visit. 'Are you in need of anything?' cry all the voices at once. 'If, on the contrary, you want some provisions,' I say in turn, 'I have brought a great deal with me. Come to my camp at only an hour's distance. There are only thirty of us.' When you appear to want nothing, every one offers you what you do want. As soon as the old women are gone, then come the young girls," (oh, oh !) "full of curiosity, and pretty underneath their colouring of Florentine bronze. Young men follow close—naturally. Then begin the rejoicings under the tent. They bring sheep, goats, dates, milk, and everything that can be agreeable to us. Strange to say, these people would never receive money from me. Nevertheless, they would perhaps have killed me had I appeared before them armed."

Another day the Viceroy said to me, "You are privileged, it seems. I had a dinner-service. It has arrived broken to bits." "If you took the precautions I take," I answered—"if you did not trust your crockery to persons who take no care of it—it would be otherwise." So the Viceroy, to replace the camel which usually carried my crockery, and which was tired, chose another one very lively and nearly wild, which jumped about with my plates and my dinner-service, to the great amusement of the Prince, who held his sides while viewing the disaster of the outfit he had given me himself. (Laughter.)

After a journey of three months we returned to Cairo, where everything was threatening. The English Government, by the lips of Lord Palmerston, had in Parliament made use of unpleasant expressions concerning me. He had represented me as a species of pickpocket, wishing to take the shareholders' money out of their pockets.*

* I think M. de Lesseps makes a mistake as to the date at which these expressions were used. In 1857 I find in 'Hansard' two short speeches of Lord Palmerston's, of July, of which the following are extracts :—

(General hilarity.) The alliance of France and England for the Crimean war still

Lord Palmerston in answer to Mr H. Berkeley, July 7, 1857. —"The obvious political tendency of the undertaking is to render more easy the separation of Egypt from Turkey. It is founded also on remote speculations with regard to easier access to our Indian possessions, which I need not more distinctly shadow forth, because they will be obvious to anybody who pays any attention to the subject. I can only express my surprise that M. Ferdinand de Lesseps should have reckoned so much on the credulity of English capitalists as to think that by his progress through the different commercial towns in this country he should succeed in obtaining English money for the promotion of a scheme which is every way so adverse and hostile to British interests. That scheme was launched, I believe, about fifteen years ago, as a rival to the railway from Alexandria by Cairo to Suez, which being infinitely more practicable, and likely to be more useful, obtained the pre-eminence. M. de Lesseps is a very persevering gentleman, and may have great engineering skill at his command; at all events he pursues his scheme very steadily, though I am disposed to think that probably the object he and some others of the promoters have in view will be accomplished, even if the whole of the undertaking should not be carried into execution. If my hon. friend the member for Bristol and his friends will take my advice, they will have nothing to do with the scheme in question."

On July 17, Lord Palmerston in his speech said—"I therefore think I am not much out of the way in stating this to be one of the bubble schemes which are often set on foot to induce English capitalists to embark their money upon enterprises which in the end will only leave them poorer, whoever else they may make richer."

lasted. Furnished by M. de Rothschild with recommendations, I began to hold meetings, which I continued in England, Ireland, and Scotland for twenty-two days. As an instance of the liberty of speech enjoyed across the Channel, I may say that at Liverpool, the mayor, knowing my wishes, offered me his co-operation, prepared the hall, issued advertisements at his own cost, and took the chair at the meeting.

I scarcely expected a favourable reception. But it was far otherwise. Despite the frightful jumble of English words, which I drowned in the midst of French expressions, every one applauded me, wishing to show that they understood me perfectly. I thus travelled over Ireland and Scotland, accompanied by Mr Daniel Adolphus Lange,* our representative in London, who was of the greatest service to me.

* Knighted in 1870 about the same time that M. de Lesseps received the order of the Star of India of the First Class.—H.D.W.

On arriving in a town I always called on the writers of the press. I begged them to come to my meeting. They came, and never did I give them a penny. In the evening I corrected the proofs. I took with me a thousand copies; and the next day I went to another town, where I distributed them. I begged the most important personage of the place to be good enough to preside.

There are to be found everywhere persons who love to do a service, and who, in the public interest, gracefully consent to what is asked of them. I chose a secretary to send round invitations.

Freedom of speech is in no wise shackled in England. It is, on the contrary, asserted and fostered by every one. One day, on arriving at a place, I learn that the most considerable person was a lord, who was inspecting the prison as a justice. I entered without difficulty; but when I wished to go out, I found the gates shut. (Laughter.)

Another time my candidate was presiding over a court of justice. After the first case was over, I begged him to receive me in his private room, and told him I wished to speak in public. "All the world may do so," he answered. He first wished to be excused from taking the chair, on account of his engagements; but, on my pressing him, undertook everything—the expense of summoning the meeting, room hire, and other details. It is thus things go on in England. It is easy to perceive that truth always results from discussion. The most absurd things are freely listened to, because they provoke good and useful explanations. Our high society is, to my mind, more irreconcilable than the poor of the lower class. Why not instruct them?—why prevent their learning? I once found myself at Marseilles in a heated popular assembly composed of three thousand people. I did not fear facing them and defending what they were attacking. Prosecute them, obstruct liberty of discussion, and the truth will never reach these men. We thus stimulate the fatal doctrines propagated by secret societies. (Marks of assent.) I approve teaching our children Greek and Latin; but what we must not neglect is to teach them to think wisely and to speak bravely. (Hear, hear.)

Men are generally of good faith. When they are told the truth, they listen, and renounce their errors.

My addresses having given full satisfaction, and public opinion being favourable, I had only to follow it up. I returned to Egypt and Constantinople, and employed the success of my meetings to counterbalance the effects of English diplomacy.

I succeeded only in 1858. As you see,

the first steps were long and laborious. Fancy that in the first four years I travelled ten thousand leagues every year—more than a journey round the world!

Opposition was not long in becoming less active at Constantinople. The good Turks were always saying to me—"Do what you like, only take care to come to an understanding with the Powers, that they may not be unnecessarily tormenting us."

I continued, therefore, going from Constantinople to Cairo, and *vice versa*, until the time arrived when I asked the public for the capital. I have been much reproached for this act of boldness.

The preparatory studies were in a forward state. I had planned a circular with my friends. I had even concerted the organisation. Everything was ready; but I remained at Constantinople, fearing that, in the absence of a firman, a protest might be made by the Porte. We found ourselves in a difficult situation, and our opponents did not fail to profit by it.

Nevertheless, I decided to leave for Odessa, where I was wonderfully well received, and for the chief towns of Europe. I held meetings, which excited, as in the theatre of Marseilles, transports of enthusiasm in spite of all the financiers, and even of some of my friends, who reproached me with rashness, which might compromise everything, and make success impossible. Nevertheless, I was advised to open a subscription at Monsieur de Rothschild's. I had rendered him some services while Minister at Madrid, and he was good enough to recognise them.

"If you wish it," he said, "I will open your subscriptions at my offices."

"And what will you ask me for it?" I answered, enchanted.

"Good heavens! it is plain you are not

a man of business. It is always five per cent."

"Five per cent on two hundred millions (£8,000,000); why, that makes ten millions (£400,000)! I shall hire a place for twelve hundred francs and do my own business equally well." (Approving laughter.)

Well, the Grand Central had just left the Place Vendôme. There I established my offices, and thither capital flowed in abundance.

By the advice of the Viceroy I had reserved for foreign Powers a portion of the shares. But France alone took on the whole amount 220,000, the equivalent of one hundred and ten millions.

I witnessed in the course of the subscription some curious facts full of patriotism.

Two persons wished to subscribe. One was an old bald-headed priest, doubtless an old soldier, who said to me—

"Oh, those English!"—(laughter)—"I am glad to be able to be revenged on them by taking shares in the Suez Canal."

The other who came to my office was a well-dressed man, I know not of what profession.

"I wish," said he, "to subscribe for the Railway of the Island of Sweden" (le chemin de fer de l'île de Suède).

"But," it was remarked to him, "it is not a railway, it is a canal; it is not an island, it is an isthmus; it is not in Sweden, it is at Suez."

"That's all the same to me"—(renewed laughter)—he replied; "provided it be against the English, I subscribe."

The same patriotic eagerness was found in many priests and military men.

At Grenoble a whole regiment of engineers clubbed together to have its share in a work so eminently French.

Even men of letters and retired public servants, who generally do not invest a sou in business, showed their desire to encourage our efforts.

The Comte de Rambuteau, who was blind, said to me one day—

"I have never placed a centime in any enterprise whatever; nevertheless, I have taken two of your shares."

"Those two shares give me more pleasure," I replied, "than a hundred thousand others bought by a banker, for they are a fresh proof of the sympathy of France in my undertaking." (Hear, hear.)

I will now pause for a moment. You must require rest.

Here followed a short interval, after which M. de Lesseps resumed thus:—

We now come to the second part of this

lecture. I say "*we*," because you take as much part in it as myself. But for your goodwill, I should certainly speak with less ease than I do before you. I speak to you as to friends. (Applause.)

We have reached the moment at which the Company is constituted financially. The Council of Administration sends a Commission to take possession of the land. We present ourselves with a statement addressed to the Viceroy, whom the difficulties continually raised since the formation of the Company had rendered so impatient that he would no longer listen to us, and would only grant us the shortest possible audiences. To let him know the contents of our letter, we were obliged to place it on an arm-chair, and take it back again, so that he should not appear to have received the notification of the Company being in existence. As I knew that in reality we could count on him, we always maintained an extreme reserve. We left for Cairo, and he for Upper Egypt. One day he learns that I found it necessary to go to Cairo, where he was staying. He leaves by rail at once, taking his nephew, the present Viceroy, and his brother, and presses on the train at such a pace that his brother says to him, "Monseigneur, we run more danger on the railway than with Monsieur de Lesseps." (Laughter.)

Without comparing myself to Moses, one thing used to astonish me when young, when reading the Bible. There one sees that he used to enter Pharaoh's presence, reproach, and even menace him. How does it happen, I asked, that so great a sovereign did not turn this man out of doors, or allowed him to come near him? (Renewed laughter.) This is the reason. In the East, when a prince has in his youth known any one, he cannot forbid him his threshold.

So the Viceroy adopted the course of absenting himself. For a long time, when difficulties started up on every side, nothing worried him more than speaking of the Canal. He begged me to remain several weeks without seeing him. He told every one to grant me nothing, while secretly he allowed assistance to be given me. Thus, in an encampment where we were refused water, one of our engineers could only obtain some by threatening the captain of the boat pistol in hand. Before his ministers the Viceroy was indignant at such conduct, of which, I am certain, he approved. In public he said he had withdrawn from me his friendship—that he forbade all assistance to us, &c. One day in full council he had indulged in some such tirade. Every one had left the room, when, in a corner, the Viceroy espied the governor. of the town. "What are you doing there?" he asked.

"Did you not hear my orders?" "Forgive me, sir; but your Highness gave them with such violence that they cannot possibly be your real intentions." "You have understood me," said the Viceroy. "Begone! But take care; for if you allow it to be suspected that I have authorised you to help Lesseps, you'll have to answer to me for it." (Laughter and applause.)

So the very next morning I had the audacity, at least in the eyes of the public, to inquire amongst Europeans for persons willing to enter our service. All natives had been driven from our yards. None but French remained. Our fellow-countrymen are always firm and steady at their post. Without them I should never have made the Canal, which is really the work of their mind and their energy. (Loud applause.) That day I hired, for twelve hundred francs a-day, a steamer which belonged to the

Government. I embarked on it persons of every kind to the number of two hundred. I placed myself at their head, and the police did not ask for papers.

On leaving the port I did not venture to ask for a bill of health, preferring not to bring the despotic sanitary authorities on my shoulders. Since the famous plague of Marseilles in 1750, every sort of precaution is taken to ward off a disease which occurs very seldom, and which quarantines do not stop when it is destined to come. Precautions are invented which are perfectly useless and which injure trade. (Marks of approbation.) It was thus that the first vessel of the Messageries Impériales, which arrived from India through the Canal, was detained five days at Marseilles.

At Damietta I found a sanitary officer whom I took with me. "Supposing I lose my place?" he asked. "I will give you another," I answered. (Approving laughter.) He comes with me to the governor, who, we are informed, is in bed. Well, as there is no governor, we are masters of the town. (Renewed laughter and approbation.) We take our provisions, and return on board in a boat. Some days later I inquire of the governor as to the serious illness which kept him in bed when I wanted to see him. "It was like this," he answered: "I had sent a telegraphic despatch to the Viceroy, informing him that you had collected men and provisions to be taken to Port Saïd; and I asked for his instructions." "Imbecile," replied the Viceroy, "this is not the way to write 'Saïd!' Finding the solution so little clear, to cut short every difficulty, I took to my bed." (Laughter.)

Let us now accompany from Cairo the administrative Commission charged with taking possession of the land of the isthmus.

Application was made to the chief camel-driver of Cairo for a hundred camels. He pretended not to have them. When this news was brought to me I was exhorting my companions to have patience with the Arabs. I interrupt my conversation, and going to my room find the chief camel-driver, and frighten him so terribly that he throws himself on his knees and promises all I want. I take him before the governor, and the order is given to form our caravan.

We arrived at the last village in Lower Egypt. While my companions go shooting, I am told that an officer of the Cairo police, who had been following us for several days, has seized some of our camel-drivers, and imprisoned them, with ropes round their necks.

I immediately go to him, and after having asked for his instructions, which he could not show me, I treated him before the public in such a manner as to show the population that I was much his superior. In the East one must be either the hammer or the anvil.

Our last station, before plunging into the desert, was near to Korein, on the road to Syria, where Greek philosophers, patriarchs, great conquerors, the Holy Family, and Napoleon I have passed. Some of our men ask for water and milk. They are answered that there is none. The truth was, as I knew, that the Cairo police-officer, who continued to follow us, had incited the inhabitants of the village to refuse us all provisions. I assemble the principal inhabitants of the place in my tent. At this moment we were running great danger, for it had been announced at Alexandria that we had been assassinated and massacred by the Arabs. Of this I knew nothing. Nevertheless I took care to give our visitors to understand that I was not a man to allow myself to be touched with impunity. Thereupon, after coffee, I show them a revolver, which, contrary to my habit, I had amongst my baggage, and which I was taking as a present. I placed six empty bottles in a row, at a certain distance, and with the six barrels of my revolver I break them, to the great stupefaction of our guests. "Recollect," I said to them, "that we have twenty in my band, and that I am the worst shot of all. We are going into a desert, where we shall take every black point for a gazelle." No one disturbed our journey. We accomplished it in all tranquillity. We took possession of the land, and turned the first spadeful at Port Saïd, to the great disturbance of Lord Palmerston.

On arriving at Suez, the governor of the town, accompanied by the police-officer I had brought to reason, made me excuses.

The Viceroy had promised me twenty thousand men; but in 1861 he was so tormented, there was so much animosity shown in diplomacy, that he begged me, with a certain justice, not to keep him to his engagements. I myself advised him to observe great prudence. It was then that I undertook a journey to my friends the Philistines, a population of solid and vigorous workmen. As they hold all the plains from the confines of Egypt to the mountains of Jerusalem, they have always been the terror of travellers. Nevertheless it often happens that men, like horses, are only bad because they are afraid. (Laughter.) If you appear before them armed, they will kill you for fear you should kill them. It is quite natural. I travelled on a dromedary, accompanied only by two persons. In crossing the sandhills of Katieh, about thirty or forty leagues in length, with hillocks four or five hundred feet in height, composed of extremely fine sand, we lost our way.

While riding in advance of my companions, I remarked on the side of the plain a road which appeared to me the road to Syria. I called to my comrades, who were following at a distance, to come up to me. At the sound of my voice, four men, armed with swords and pistols, leave a wood where they had lain hid, throw away their cloaks, and rush towards us.

I was on an eminence. "Well, my friends," I asked them, "why do you run so fast up to us?"

"We thought," they said, "that you had lost your way, and we came to succour you; because if the night surprised you in these sandhills you would run great danger."

Perhaps these men were there to rob travellers. (Laughter.) But they thought me in danger, and they came to my assistance, as their religion obliges them. This may serve as a study of the human heart.

When I met groups of Arabs, I advanced to them alone. I saluted them in the name of God. Far from doing me harm, they would invite me to their tents, where I found the best hospitality. The women dried my clothes, gave me coffee, &c. In each village I circulated in great numbers a proclamation I had printed, calling on the population to work. I told them that, till then, they had lived like tigers; and that if they wished it they could make much more money by working on the isthmus, and would incur less danger than in wandering on the highroads, at the risk of catching rheumatism or bullets. You have no idea of the ovation which these people gave me during my whole journey. On the frontier of Egypt, at El Arish, the inhabitants carried me on their shoulders to the top of the citadel, where the governor entertained me.

The chiefs of the town accompanied me to the frontier of Egypt and Syria, singing psalms and canticles. These details interrupt my narrative, but the attention with which you listen encourages me to continue. (Go on.) At the time of the war in Syria, in 1834, Ibrahim Pacha had cause of complaint against the population of Bethlehem, which is Catholic. He therefore sent to the galleys all the inhabitants fit to bear arms—four hundred young men—and, doubtless as ringleaders, twelve old men.

As president of the sanitary commission, I observed at each of my visits of inspection those twelve aged men, and the four hundred young ones, who were intoning hymns in honour of France. I asked them what they required, and what they had done. "We have been reduced to slavery," they said, "because of our alliance with the chief Abou Gosh." This was a chief who commanded the pass where David killed Goliath. Abou Gosh, the descendant of an ancient family, dating from 1100, was then resisting with all his might the domination of the Turks over his fellow-countrymen. I presented myself to Mehemet Ali. I interceded unofficially with him for these unhappy Catholics. I begged him to restore them to their families. Mehemet Ali answered, "I cánnot promise you all you desire, and what I desire myself. I fear annoying my son Ibrahim by releasing all the prisoners whom he wished to punish for their revolt. But be easy. Every week I will place five of them at your disposal."

No sooner was the news known in Bethlehem than my door was besieged by the wives and relatives of those who were detained in the galleys. I could not leave the house without being, like the great men of antiquity, surrounded by an unhappy crowd which came to solicit my protection. They pressed on me on every side, and tore my clothes. Meanwhile Ibrahim Pacha continued the course of his victories on Mount Taurus, and I considered that a little more generosity could be shown the Bethlehemites without offending him.

In this state of things it occurred to me one day to go to Mehemet Ali with my clothes in rags. "What is the matter?" asked the Viceroy. "It is your fault," I answered; "and I do not know that it can last. So long as you do not release my *protégés*, who are kept at the galleys, it will be the same; and I shall never get to the end of my troubles if you only release five prisoners a-week." At length the Viceroy yielded to my prayers, and allowed all those good people to return to their country.

Thirty years later, in the journey of which I am now speaking, from the first day of my arrival in Jerusalem, old men in red robes came to greet me and thank me, saying, "It was thou who saved us formerly by turning away the vengeance of Ibrahim Pacha. Be blessed." Though charmed with this pleasant meeting, I was rather vexed to see that men of my own age were already so old. (Smiles.) There were then at Jerusalem a hundred French cavalry and fifty officers of the staff accompanying General Ducros, who belonged to the French Expeditionary Corps. As they had come to assist at the Easter festivities, I invited them to accompany me to Bethlehem.

Since the crusades, such a sight had not been seen as French cavalry deploying along the mountains of Jerusalem, with their trumpets at their head. We found on our road, stationed at regular distances, first young men, then old men, who gradually increased our *cortège*. On reaching Bethlehem the town was in a *fête*. The women burnt incense under the nostrils of my horse, and, as is the custom, slaughtered lambs in the streets. From the windows and roofs they sang our praises, as is the Eastern habit, and our path was strewn with verdure and flowers. The French officers did not seek to hide their emotion. On arriving at the grotto of the Nativity, an old man coming forward presented to me a child. "Here," he said, "is a son of those whom you saved." (Boisterous applause.)

I thank you, gentlemen. Believe me, if I tell you these things it is not to provoke your cheers; it is because they were the beginning of that universal impulse and enthusiasm which time has not weakened, and which have achieved this great work. (Renewed applause.)

Ismail Pacha, on acceding to power in 1863, proved most loyal towards me. This prince, like his father, is a good administrator, and showed himself anxious to set in order the condition of the Company.

Against this view may be objected the theatres and the actors on whose account he has of late incurred great expense. Yet this is one means of civilisation. One may civilise by science, but one can also civilise by pleasure. (Hear, hear.) The Viceroy wishes at any cost to regenerate the morals of his country. He wishes to reform the harems, which are a source of intellectual and moral abasement. (Approbation.) He wishes women to play their part in society. He has already reserved for them boxes at the theatres, from which I trust he will some day remove the gilded gratings.

I am grateful to him, in the name of French civilisation, for having appealed to

France for the amusement and instruction of his subjects. He has understood that women in society is the first element of progress.

The Viceroy thoroughly feels that all improvement among the Mussulmans is impeded by the unjust inequality existing between man and woman.* In the East the world marches only on one leg. That is why it is so behindhand. (Hear, hear.)

One day I was riding with the governor of Suez, an intelligent man, brought up in Turkey.

"How is it," he said, sadly, "that we remain for ever behind you? I have friends who have studied in France, in England, and in Germany. Why, so soon as they return to the East, do they do like every one else?" At that moment there passed, mounted on

* This was also the view of Fuse Pacha, one of the greatest of Turkish statesmen.—H. D. W.

horseback, the young daughter of the English consul. "When your wives and daughters shall thus gallop at your sides," I answered him, "you will be a civilised people."

I said the same thing to the Viceroy, who seemed much struck with it. He desires to make use of the means which have civilised Christians; for the Mussulman religion is not opposed to progress. A verse of the Koran says—

"He who obstinately wishes to do for ever what his father did before him, merits the flames of hell."

Ismail acceded to power in 1863 with the same difficulties as his predecessor, arising from English opposition. But he succeeded in overcoming them assisted by the arbitration of the Emperor, which he himself invoked.

We at length emerged from our political difficulties and obtained the Sultan's firman.

Then with the assistance of Messieurs Borel and Lavalley, and thanks to their gigantic inventions, we pressed on the works with an activity which, it may be said, has had no precedent in the history of industry.

Our dredging-machines, of which the ducts were one and a half times as long as the column in the Place Vendôme, carried off from two to three thousand cubic metres a-day; and as we had sixty of them, we succeeded in extracting monthly as much as two million cubic metres (about 2,763,000 cubic yards).*

This is a quantity of which no person can form an exact idea. Let us try to realise it by comparison. Two million of cubic metres would cover the whole of the Place Vendôme and would reach an elevation of

* For an account of this machinery see an interesting article in the 'Fortnightly Review' of January 1869, by Captain Clerk.—H. D. W.

five houses placed one on the top of the other.

Two millions of cubic metres would cover the causeway of the Champs Elysées as high as the trees, between the obelisk and the Arc de Triomphe; or again, the whole of the Boulevard, from the Madeleine to the Bastile, would be covered up to the first floor of the houses. (Surprise.)

This is what we carried away in a month. It took four months for the 400,000 cubic metres of the Trocadero, while we dug out two millions in one month. Let us do justice, gentlemen, to the men of science and of courage who executed this immense labour. They have deserved well of their country and of civilisation.

Some months since we had to announce to our general meeting that the Canal would be opened on the 17th November. So indeed it was, but not without difficulty, not without terrible emotions. I never have seen so clearly how near is failure to triumph; but, at the same time, that triumph belongs to him who, marching onward, places his confidence in God and man. (Loud applause.)

Fifteen days before the inauguration of the Canal, the engineers came to tell me that, between two soundings, taken at distances of 150 metres, by means of square shafts, holding twelve men, a hard rock had been discovered, which broke the buckets of our dredgers. We have been blamed for not perceiving it sooner. Was it possible to take soundings at shorter distances in a length of 164 kilometres (about 102 miles)? At this sad news I hasten to the place pointed out. There we found a boulder rising five metres above the bottom of the Canal, and leaving only three metres of water. What was to be done? Every one began by declaring that there was nothing to be done. In the first place, I cried, "Go and get powder at Cairo—powder in masses—and then, if we cannot blow up the rock, we will blow up ourselves." (Laughter and applause.)

The sovereigns were on their road to the rendezvous. All the fleets of the world had been bidden, and were about to arrive. It was necessary at any price to be in a position to receive them. The intelligence and energy of our workmen saved us. Not a minute was lost, and all the ships were able to pass. (Applause.)

Enchanted with this result, the Viceroy came to me and asked me to make the necessary arrangements for receiving the sovereigns and the foreigners to the number of 6000, whom we were to shelter and feed. Sheds were constructed in a few days to hold 600 persons, with tables constantly replenished and served.

The Viceroy had brought over 500 cooks and 1000 servants from Trieste, Genoa, Leghorn, and Marseilles. There was also, opposite the Sweet-Water Canal and Lake Timsah, a village of 25,000 Arabs, who were likewise affording hospitality under their tents. All these preparations were ready when, on the 15th, as I was about to leave for Port Saïd, at nine o'clock in the evening, I heard a sound of petards and rockets bursting. It was the fireworks which had been brought for the *fêtes*, and which, having arrived too late by the railway, it had been impossible to convey, as I had wished, to the sandhills outside Ismaïlia. They had been placed in the timber-yard in the middle of the town, which narrowly escaped becoming entirely a prey to the flames. Two thousand troops came opportunely, and the town was saved — thanks to the system always employed at Constantinople, and which consists in unceasingly pouring water on the walls and roofs of the neighbouring houses.

Despite our efforts the wall became heated all round to such an extraordinary temperature that it was threatening to spread the fire, when I was told that underground in the yard there lay buried in the sand a large quantity of gunpowder. I begged that nothing might be said, and directed all the pumps that way. Fortunately the wind fell altogether, and the town was saved.

On the 16th of November, 160 vessels had arrived. The next morning the prayers both of Mussulmans and Christians were to be celebrated. Two similar platforms had been prepared to receive two altars. A third platform was destined for the sovereigns and the distinguished guests.

The different arrangements were completed when suddenly a high sea covers the beach with water, and surrounds the platforms. We did not know how to remedy this : at length we managed to form with sand a free dry space round the tribunes.

These were thus surrounded by water, and it was a magic sight to see the guests, as they arrived, cross this unexpected lake.

It was the first time that the Christian altar and the Mussulman altar had been placed side by side, and that the two clergies officiated together.

Orders had been issued to despatch on the morning of the 17th the fleet of inauguration.

On the evening of the 16th, after receiving the Empress and the foreigners, I was making arrangements with the captain of the port—a very distinguished naval officer, M. Pointel, whom death has since taken from us. We had settled everything, when at midnight we learn that an Egyptian frigate has run aground thirty kilometres from Port Saïd, in the middle of the water—that is to say, that she had run on one of the banks, and, lying across the Canal, was barring the passage. I at once collected the means necessary for getting her off. A steamer was sent off with men and appliances for the operation. They return at half-past two in the morning, saying it is impossible to move the frigate. Gentlemen, one must have confidence in this world. Without it nothing can be done. (Hear, hear.) I did not wish in any way to change the next day's programme. Logically I was wrong, but the results proved me right. (Renewed applause.) We must not be *doctrinaires.* It answers neither in business nor politics. (Hear, hear, and renewed applause.)

At three in the morning, the Viceroy, who had left for Ismaïlia to receive the sovereigns and princes, hearing of the grounding of the frigate, returned in all haste. On passing he had made some useless efforts to dislodge her. He sent for me on board his boat, and I found him in great anxiety, for our minutes were already numbered. If we had adjourned the opening even to the next day, what would have been said? Despatches by orders from Paris were already publishing that all was lost.

Powerful assistance was placed at the disposal of the Prince, who took with him a thousand seamen of his squadron. We agreed that there were three methods to be employed: either to endeavour to bring back the vessel to the middle of the channel, or to fix it to the banks; and if these two means fail, there was a third. We look into each other's eyes. "Blow it up!" cried the Prince. "Yes, yes; that's it. It will be magnificent." And I embraced him.

(Salvos of applause.) "But at least," added the Khedive, smiling; "you will wait till I have taken away my frigate, and that I have announced to you that the passage is free." I would not even grant him this respite. (Laughter.) The next morning I went on board the Aigle, without mentioning the accident to any one, as you may well believe.

The fleet started, and it was only five minutes before arriving at the site of the accident that an Egyptian admiral, sailing on a little steamer, signalled to me that the Canal was free. (Bravo!) On arriving at Kantara, which is thirty-four kilometres from Port Saïd, the Latif, dressed in flags, saluted us with her guns, and every one was charmed with the attention which had thus placed a large frigate on the passage of the fleet of inauguration. (Cheers and laughter.) On arriving at Ismaïlia, the

Empress told me that during the whole journey she had felt as though a circle of fire were round her head, because every moment she thought she saw the Aigle stop short, the honour of the French flag compromised, and the fruit of our labours lost. (Sensation.) Suffocated by emotion, she was obliged to leave the table, and we overheard her sobs—sobs which do her honour, for it was French patriotism overflowing from her heart. (Applause.)

We passed without difficulty the rock of Serapium; and what gave me great pleasure as we were skirting it, was that the workmen near it, after looking to see if we touched the bottom of the Canal, expressed their transports of joy by a gesture which no expression can render. (Here M. de Lesseps, by imitating the action of the workmen, brings down the applause of all the hall.)

It must be said that from the beginning of the work there was not a tent-keeper who did not consider himself an agent of civilisation. This it was made us succeed. (Hear, hear.)

The passage was effected marvellously. One hundred and thirty ships inaugurated the opening of the Canal, and since that day there has been no interruption to the traffic. Henceforward the Canal is opened to all ships, whatever their draught of water.

Steam navigation sees opening out before it not only Arabia, China, Cochin China, Japan, and the Philippine Islands, but the eastern coast of Africa, which, by its streams and rivers, offers such marvellous resources for commerce. Very rich coal-mines have been found there. From Japan to San Francisco, multitudes of archipelagoes, scattered over two thousand leagues of the Pacific Ocean, call for the colonisation, not of Governments, but of individual initiative.

After the old example of our younger sons who conquered Canada, Louisiana, and India, let the youth of our day, instead of living in idleness, or following up careers which lead to no good, go and fertilise new "Isles of France."

Let nothing discourage them. The spirit of initiative and perseverance belongs to our nation more than to any other. (Applause.)

Ladies and Gentlemen, I thank you for your kindness, and hope that you will give us your wishes that the Canal may succeed for its shareholders as it has succeeded for the science and honour of France.

(M. de Lesseps is greeted by redoubled applause, and the meeting separates, deeply gratified.)

Printed in the United States
By Bookmasters